GLOBAL WARMING:

WHY YOU SHOULD BE AFRAID

GLOBAL WARMING:

WHY YOU SHOULD BE AFRAID

JOSEPH PATRICK NOONAN

iUniverse, Inc.
New York   Bloomington

iUniverse books may be ordered through booksellers or by contacting:

iUniverse
1663 Liberty Drive
Bloomington, IN 47403
www.iuniverse.com
1-800-Authors (1-800-288-4677)

Because of the dynamic nature of the Internet, any Web addresses or links contained in
this book may have changed since publication and may no longer be valid. The views
expressed in this work are solely those of the author and do not necessarily reflect the
views of the publisher, and the publisher hereby disclaims any responsibility for them.

ISBN: 978-1-4401-3828-7 (sc)
ISBN: 978-1-4401-3829-4 (ebook)

Printed in the United States of America

iUniverse rev. date: 4/23/2009

Dedicated to my father, Francis Joseph Noonan
1919 – 1999

Contents

PREFACE

I am 54 years old, I think a typical middle class American. I read and watch TV like most. How I am different is, I like politics, history and science. Issues, such as global warming, intrigue me. My goals are simple. Provide for my family now and in the future. I have never been poor in the sense that I always had the essentials, but I have never been rich either. My wealth is my family and friends and money has always been secondary to me. My inspiration for this book is my wife of almost 20 years. She and I like all couples have our ups and downs. We both travel for business and are very busy at times. I, like many, have squandered my life by not prioritizing my time for my family first always so busy thinking tomorrow is another day. That tomorrow came and is gone but hopefully I still have a few tomorrows left. I started to write this book because I needed a hobby that would occupy my free time in a productive fashion. My hope was and is to reinvent myself so as to get in touch with my priorities and hopefully reinvigorate my life. I could not have accomplished this without the support I received from my wife and family. I care deeply about the future and particularly our (Americans) future and hope the thoughts and opinions I have expressed are a source of intelligent debate.

INTRODUCTION

As I learned the book writing process I realized putting thoughts on paper is much more difficult than it appears. The strangest thing that happened was I found that I have something to say. I believe many people do but never get the chance. As you all know, political, and religious conversations at parties or family or social gatherings tend to start arguments and ruin any enjoyment for those who are just trying to relax. So for most of us there is no place for intense intelligent conversation. I am not important enough to be taken serious but I came to realize that is a big part of our society's problem. We tend to be sheep. Instead of engaging each other, we depend on those on television to form our opinions. It seems to me there is plenty (too much) commentary on television and less and less factual reporting. It is time to get informed and form our own opinions. I personally have never had a deficiency in opinions, just ask anyone who knows me.

I dedicated this book to my father. We rarely talked and probably agreed with each other even less often. I can not give you his opinions but I can tell you my father was way ahead of his time. He spoke of climate change as far back as I can recall. In the late 1960's he had formed the opinion that pollution (greenhouse gases) would lead to another ice age. The jury is still out. But looking back I realized that (although he was more intelligent than most) climate change theories are not new and in fact all the proposed solutions existed then. "No new technology" has been developed to suggest that we can solve our climate or energy problems today any better than we could have solved them then. He spoke of ethanol powered vehicles that already existed. He spoke of natural gas as a substitute for gasoline. He realized that the problem then, as it is now, is the political power struggle between two extreme factions and he said so. Another thing about my dad is, unlike some climate pundits who are wealthy, he was no hypocrite. A World War II veteran, graduate of West Point, self employed most of his life who worked and lived to support his eight children and wife (my mom). In the late 1960's, much to my horror, he experimented with composting bio-waste and a vegetable garden in our backyard. We lived in Baltimore city and we had to bear some ridicule from neighbors and friends. That was not a problem because if he taught us anything, it was not to be too concerned with other peoples' opinions especially when they were often poorly informed. (Poorly informed opinions are similar to a part of the human anatomy and everyone has at least one) When he retired he purchased a piece of property and built an energy efficient home featured on the front cover. What you might call a well insulated shack but very comfortable. He heated with wood harvested from his property. He for a while grew some of his own vegetables and if he had chosen, he could have hunted animals for meat or had livestock for meat, milk and eggs. This is not an option for most people and is not the point. He valued the environment and lived according to his beliefs even though he had little money and quite often was called a nut. In more ways than I can articulate, he was a hero!

All of a sudden I thought that maybe a common man can make a difference. What I mean is, if anyone will listen, maybe I can inject a few ideas that get past the partisan fighting and we can start to address the problems of our world. Maybe I can offer a perspective that seems to be lost by the powers that be. I started to research the issue when I realized the data was insignificant because global warming has become such an emotional and politicized issue that I think no one is ever going to really prove their point of view or be able to be willing to offer real viable solutions. I'm not promoting global warming, like some, and I won't deny its existence either. I think, like any real problem, if you look at the situation and honestly want to do the right thing, it is not very serious or at the very least, it can be addressed without destroying our way of life, culture or prosperity. Too many people are suggesting a radical change in the American way of life. This will not fly. We need solutions that protect and preserve our way of life!

So I started to think about a solution to global warming and thought, what if? I might seem to be trying to criticize the political and social positions of those we trust to manage our society but no matter what your political position, there is plenty of criticism to go around. I first became aware of energy and climate issues way back in junior high school, mostly because of my dad. I suggested a machine that generated more power than it consumed and I was immediately told how stupid an idea that was and that a perpetual motion machine was scientifically impossible. I talk about that incident because it shows how often ideas stupid or not are not explained so as to promote knowledge and independent thinking. Too often, ridicule is all you will get for thinking. Back in the mid 1970's my first home was heated with oil and I took a financial beating that I will never forget. At the time I worked as an auto mechanic and remember the beginning of the emission control era. I remember Carbon Monoxide being one of the gases emitted that had to be eliminated from auto exhaust or at least the volume dramatically reduced and no mention of Carbon Dioxide. We'll get to that later. I am not a scientist not even close! I am not a politician, and I take offense to any such notion. I am not even a scholar so you can say I don't know what I'm talking about. I am not trying to be humble or insignificant either. I'm just an average guy, who's only right to have an opinion is the fact that I have one. I'll try not to be too one sided and I am not trying to perpetuate global warming related arguments. I am, however, quite conservative and believe true conservatism is all but dead. Obviously, there are a lot of issues.

Yes, it is possible that I am insane. I'll leave that to you. Anyone who thinks that there are real and practical solutions to any problem in our current social and political climate probably is. Is my rubber room ready? I do need a place to retire soon. I am by nature a technician. I am one of those people who are cut out to be a certain thing, a technician. All that means is that I look at any problem as a challenge and evaluate it by breaking it down into solvable components. Then you fix each item, one at a time, until the problem is solved. It comes down to applying simple logic and avoiding emotional distraction.

The most significant conclusion I came to is a little more profound. The global warming issue may be one of the best things that happened to this country in a long time. It could be the thing that unites us if we let it! Or to put it another way, if we can solve the pieces one at a time to our advantage, in spite of our differences and accept the ones that are out of our control and *together* we can build a better society. Neither side can do it alone! I'll try to explain.

Chapter 1: What Is Global Warming And Is It Real?

The Science: First off, global warming is real! The science however is incomplete and mired in extremist theories not facts. At least at the moment, the planet is warming up. The global warming advocates believe the cause is a dramatic increase in greenhouse gases in the atmosphere. All data supports both but is inconclusive. They also believe that the activity of people, particularly industrialization, is the source of the damaging greenhouse gases. Since the industrial revolution started the amount of greenhouse gasses has increased dramatically and the average temperatures have been climbing. That would support the notion that they are related.

Greenhouse gases are believed to allow the sun's rays in through the atmosphere but not allow them to escape, causing the earths temperature to rise. Simple enough! There is another theory. The greenhouse gases will block the sun's rays and cause the planet to cool down and there is some recent evidence the planet is cooling down. There is one other theory that the planet will heat up, change the climate patterns that keep us warm and plunge us into another ice age. No one knows. Greenhouse gases include many, but they say Carbon Dioxide (CO2) is the most prevalent. CO2 is good for plants. Plants use CO2 and give off oxygen which is good for us. I was always under the impression that fossil fuels (particularly gasoline and natural gas) give off Carbon Monoxide (CO). You know how important they say it is to get a CO alarm in your home if you heat with natural gas or oil. CO is a silent odorless killer! A simple leak in your flew can put your health and life at risk. Cars and industry are blamed for most greenhouse gases but I have not seen any reference to CO. Maybe once CO is emitted into the air it bonds with another oxygen molecule and makes CO2? Can nature do that? I don't know I am not a chemist. I thought CO was also a greenhouse gas. I must be wrong because there is no reference to CO in any of the data I have read. But it has no effect on the global warming issue because, CO or CO2, they must both be greenhouse gases or at least pollutants and eliminating fossil fuels will reduce both.

Anyway, the issue is not whether or not the climate is changing, because they forget to tell you the earth's climate has and always will be in a constant state of change! Another thing they forget to tell you is, there is no concrete evidence, warming or cooling, whatever the cause, that the climate change they claim is happening is even a bad thing. They simply don't know. What I mean is a warmer planet might be exactly what the people of this planet need! It may lead to even more prosperity for more people like has been the case for the last two hundred years or so. The issue is whether or not our use of fossil fuels and the discharge of pollutants are causing extreme climate change. A sudden change would definitely be very bad because it takes time to adapt while gradual change is and always has a real issue for mankind and nothing to be afraid of.

The science is incomplete and certainly very confusing to a lay person, but the global warming issue is very real and very dangerous regardless of climate change. Many say there is a consensus in the scientific community about global warming and that is good enough. I don't think so! Once there was a scientific consensus that the world was flat, but it did not make it true and just recently one of Einstein's theories that the scientific consensus judged wrong (something to do with relativity or the universe or something) turned out to be correct and until recently the consensus was life could not exist on this planet without sunlight. Also wrong! Scientific consensus often turns out to be wrong.

The type and source of the energy we use is the only important issue and polluting and waste should never be tolerated. Stop waste and pollution and find renewable sources of energy with a proper economic and ecological balance and both "global warming the climate change" and "global warming the issue" go away!

The Marketing: The extremist believers (I call global doomers) are trying to convince us that we are causing irreversible and dangerous climate change. I am not trying to dispute that. The scare tactics are wrong but have been very effective. Check out former Vice President Al Gore's book *An Inconvenient truth*. It is one of the most effective brochures I have ever seen but I have trouble qualifying it a book. The reason I call it a brochure is, it states theories as fact, and has pictures that would scare anyone who does not think for themselves. Don't be afraid. Look at the photo's of ice melting that imply that this is abnormal! In fact, there is evidence of a recent increase in polar ice. We don't know what is normal over long periods of time. (Astronomical and geological time are both far greater than modern history). I like the map of the coast lines where many areas are under water. The picture can't be real because it has not happened in modern history (modern history you must agree includes the development of photography). All these situations are extreme, but certainly possible. But remember, they are fake! They are theory not fact and scientific consensus does not change that! The brochure however has been very effective and more and more people are sold. It is a sales brochure! Its function is to sell you on the authors' point of view or product. And how about that train commercial! The middle aged guy says "30 years" "why should that concern me" then steps aside to expose the little girl in the trains' path. WOW! That should scare you. If you did not see it, do not worry it only promoted fear not facts! The very effective scare tactics are one of the biggest things to be afraid of. Not the possibility of climate change, not the possible effects, but the fact that more and more people are getting scared and therefore more irrational. Clear thinking is already the first casualty of global warming!

The Politics: Political and social power appears to me to be the goal of the global doomers. What I mean is they are the same people that support social engineering through government control. More socialism to tell us all how to live and if they are successful our way of life is in jeopardy. If they get enough people in a panic, they will get their support and the power to control everything we are about. One example will be government regulations on where you can live, work or how many children you can have. What does that have to do with global warming?

We now see pressure on the government to do something. And they will!

Probably the wrong thing as they usually do. The political solutions of the past thirty years have made the situation worse not better. The politicians are now using climate change to attack each others positions and the power struggle continues and you an I will only be victims. They believe they will loose their power if they work with the opposition. Each would try to take credit for good and put blame for bad on the other. We call this partisanship but I call it "not doing your job"!

The global doomers are not entirely wrong, nor are they all bad. Some must truly believe. Some are good people with real concerns that must be addressed. Yes there are problems that must be solved, climate change or not, global warming is real! Not the theory of climate change. Not the effects on people! That's all speculation! No, what is real about global warming is the issue itself. It is becoming the problem. The issue, not the climate, is going to affect the way we live. The social and political fights are going to evolve into change that may be irreversible if we don't start solving the real problem. The real problem of course is how we use resources and energy and you will find no solutions in either side's rhetoric. Rhetoric is always what you here and ideas are always missing!

CHAPTER 2:   ENERGY

Our use and particularly the type of fuels we use to produce our energy are at the top of the list of concerns. We desperately need a change, global warming or not! I think this is one thing we all agree on. The beauty is, if we produce clean domestic energy and therefore eliminate our contribution to global warming, we end an issue that divides us!

Electricity: Most of our electrical power in the U.S. comes from burning coal. Some comes from burning oil and natural gas all are big contributors to greenhouse gas emissions. Coal is going to be a necessary evil for a long time. Clean coal technology is being developed and should continue even though it can never be totally carbon free. The good news, however, is coal is all American and we have plenty for centuries to come. It provides good jobs and contributes to the economic stability of many communities. They need jobs and financial security and coal can provide that. So don't promote abandoning coal, but if you are really concerned with global warming, then support higher prices for electricity to put cleaner coal burning plants in your communities. Higher electric prices are a political nightmare for some and something the working class can not afford to support. And promoting natural gas as a replacement for coal to produce electricity would increase the demand and therefore raise the price to those who heat with it and only shift the problem to another group of people. We don't need a higher cost of living for anyone not to mention that natural gas is also a fossil fuel and will do little to stop global warming. Natural gas, however, can help with other parts of the problem.

Nuclear power provides about 20 percent of our electric power and does not emit greenhouse gases. It has its own environmental hazards (what to do with radio active waste is one major concern) but when properly safeguarded, has a tremendously good safety record. The French, for example, get 80 percent of their electricity from nuclear power with no accidents to date. (What bothers me about that is, I would never think the French are better than us at anything. In this case and there might be some other examples – wine is not one of them– they are a lot better and wiser than us about nuclear power). But, like higher prices for clean coal, no politician is willing to try to convince people nuclear power is safe. And even those who support nuclear power plants do not want to live near one. Although I support a dramatic increase in the use of nuclear power and we should continue to fight for more nuclear development, I don't see a social / political solution for its detractors. Good luck with nuclear power. Therefore, in the U.S. clean coal has to be a big part of our electric power future because it is abundant and it is American and although wind, solar and tidal power can be supplemental they will not replace fossil fuels for a long time if ever! I think it is comical and mostly ironic that nuclear energy (the demon of power to doomers in general) is *green*!

Cars: As you know, almost all of our modes of transportation are fueled by oil. Jet fuel comes from oil. Diesel fuel (the same thing) comes from oil. Gasoline comes from oil. Cars, trucks and planes are considered the biggest contributors to green house gas emissions. True or not they do seem to get the most attention and blame. It does not matter because we all know the age of oil for transportation fuels is over. Not because of global warming, but because we import more and more oil every year and the cost fluctuates for a variety of reasons constantly hurting our economic stability. So we can all agree, no matter what you believe about global warming, oil as a fuel has to go. The good news is a viable clean replacement will be supported by all factions of the issue and therefore the only issue is how and when we find that replacement.

Home Heating: I do not envy anyone who heats their house with oil. As I said earlier, it has already hurt me at a time when I was not prepared for financial stress. How can anyone budget their money or plan their financial future when the price of oil can vary dramatically from season to season. The truth is, it always goes up! I hear people say "thank God oil prices dropped below three dollars a gallon". I remember being nearly bankrupt because oil was eighty cents a gallon at a time when I could hardly put food on the table! If you are a family that lives from pay to pay, as many do, the price of heating oil can force you to make tough choices. We could talk about how people must be more responsible (you know florescent lights and set your thermostat lower) but I think we will get nowhere. Simply put, some folks have very limited resources and do their best while constantly being plagued with rising energy costs, among other things, that hold them down. I know many of them believe that climate change is a problem, but I would suggest it is not foremost on their minds. What do you think? The one thing that is certain, the future of alternative fuels must include a home heating alternative (I think cheap heating fuel is doable! – read on). And climate change, once again, is not the problem. Oil as a fuel has to go!

Many industries may never be able to get away from fossil fuels because of the nature of the products they make. I don't care and neither should you. A total elimination of fossil fuels such as oil, coal and natural gas can not happen. The issue must be the proper use of the proper energy, and for that matter all resources, for the betterment of all people. Attacking any industry, such as "Big Oil", is stupid and will only put one group at odds with another.

Chapter 3:   The Problem

Industrialization: The volume of greenhouse gases has increased exponentially since the industrial revolution began. At the same time weather data suggests a constant warming period that also appears to have increased exponentially. So the two must be related, right? The statistics say 20 percent of greenhouse gases are produced by manufacturing and 30 percent comes from our cars. I don't know. I do know that large industry is the only entity that can provide the volume of food, clothing and all the other things we need to survive.

We spent almost one hundred years ignoring the environment. We went from family farms, small communities and a simple way of life to motorized transportation, large metropolises and a throw away society almost over night all from technology and our ability to use it! The result was a very high standard of living and eventually an awareness that resources were not endless and the environment had to be considered in our future advancements. We have to remember that problems like landfills, toxic waste and air pollution did not exist one hundred and fifty years ago. Technology is the thing we have over all other forms of life on this planet it is the reason we are successful and the cause of our problems. With over six billion people on the planet there is no longer a "Natural" future for mankind. Industry, with proper concerns for people and the environment, is not the problem but the solution! We have proven that with technology we are capable of almost anything, even our own demise. The proper use of technology and clear thinking is and must be how we secure an even better future.

Farming and Domesticated animals: Farms use many resources that contribute to the problem. Fertilizers are one and fuel to run their machines is certainly another. Fuels are the same issue as auto and diesel and will be solved the same way.

Fertilizers, however, are a different story. Many fertilizers are petro chemicals or to put it simply, they are produced from oil. They will never be totally eliminated, but the industry is constantly developing viable alternatives to them and the reason is obvious. The use of oil based products is always subject to fluctuations in cost and supply. The industry has the same problem as we have with gas for our cars or heating oil for our homes. They have for a long time realized that any viable alternative will help them keep there costs down and the industry healthy. They may or may not be concerned with global warming or pollution in general, but their future depends on finding renewable substitutes wherever possible. We have already learned the hard way about proper farming techniques (the dust bowl of the 1930's is one good example) and, with some government oversight and industry self interests, farming will continue to evolve and become more viable and an even more important part of all mankind's future.

Domesticated animals, however, are different. They, like all animals, excrete waste. I don't have to explain what waste is but all bio-waste contributes to the volume of greenhouse gases put into the atmosphere. All animals, including people, exhale carbon dioxide (CO2). The statistic I heard is domesticated animals emit 18 percent of all greenhouse gases! Please do not let anyone tell you we can do away with farm animals. So many products and certainly a lot of food come from domesticated animals. Don't join with anyone that will want to mandate or tax the livestock industry to reduce greenhouse gases. The food supply we enjoy can not be damaged over global doomism. Plentiful, inexpensive food is fundamental to our survival and prosperity. I think the future of farming will include the capture and storage of many greenhouse gases to be used as a resource. I don't know about the technology, but I think the research is already happening and again the future of people depends on a common sense approach to issues such as this not the destruction of an industry through taxes or politically motivated government mandates.

Population: If industry contributes 20 percent, cars 30 percent and farm animals 18 percent that is 68 percent of the CO_2 being put into the atmosphere today! Nature provides a certain amount of carbon dioxide and other greenhouse gases to keep the planet reasonably warm. At this point in time I can not figure out how much of the CO_2 content is necessary to prevent climate change or to put it another way "keep our warm fuzzy times". Obviously, not enough greenhouse gases would have a similar effect as too much. So where does the other 32 percent of greenhouse gases come from? Some has to be natural (whatever that is)! The rest comes from people! Yes, we exhale CO_2 and excrete other types of waste that contribute to global warming. And that does not account for our industrial waste. I'm not talking about factories again. I'm talking about the stuff you throw out every week. It has to decay and therefore excrete greenhouse gases. So for the sake of argument, let's assume that people currently contribute 20 percent of greenhouse gases. The remaining 12 percent must be natural? That would support global doomism and prove we are to blame! I guess if you are a global doomer you should just try to breathe less. Less exercise, for example, would allow you to breathe less and put out less greenhouse gases. How about a campaign to "Drive more, Breathe less" how ridiculous! Much too ridiculous to discuss seriously but lets' hope it keeps the "doomers" occupied.

The earth's population includes all plants and all animals! All consuming recourses to live! There must be a natural balance, but not enough CO_2 and all the plants die. That includes, by the way, farm crops. I don't know but let's talk about people.

In 1900 there were 1.2 billion people on the planet and by 1950 the population had more than doubled to 2.5 billion. In the year 2008 there are 6.7 billion people on the earth. In about one hundred years the human population has grown more than 4 times! The United Nations predicts the population of the planet will peek at 9.2 billion by 2075. Their math is based on a recent decrease in population growth so unless something changes (and it always does) there will be at least 3 billion plus more people on the planet by the end of this century. Based on the last century, I think they are very naive! Either way, there will be a whole lot more people around by the year 2100, they will be living longer (at least in the U.S.), and need energy, food and other resources.

In 1920 the life expectancy in the U.S. was 54 years of age. Now it is about 80 years of age. And the baby boomers (the largest group of people in this country) may live to an average age of 100 years! The reason for this is a tremendous increase in medical and for that matter all technology. The very thing global doomers complain about is the reason we can live longer, better lives. Stop complaining! Hey Global doomers do your part and die young, otherwise, except the fact that you are part of the problem! The rest of us, however, can not stop, we must go on.

The increase in population happened in conjunction with the exponential increase in greenhouse gases. Again carbon dioxide (CO_2) is the most significant. We exhale CO_2 and so do all our farm animals as well as all wild animals. (Why then would global doomers, as I do, support wild life conservation?). So we will breath out more carbon into the atmosphere and we will need even more energy. Global warming, if it is caused by people, will get worse and eliminating industrial pollution will not stop the continuing increase in greenhouse gases only slow it down so global warming (the climate change) is a reality we must adapt to, not try to eliminate!

The CO_2 content in the atmosphere is now about 380 parts per million (ppm). According to ice core samples, going back 600,000 years, it has never been higher than 300ppm. That would then be the number to try to achieve. If the whole world eliminated the 68 percent or so that comes from industrialization and domesticated animals, there will still be the other 32 percent. Of the 32 percent left a portion has to be natural with one billion people. Therefore, the five billion or so more people will still contribute significantly more CO_2 to the atmosphere than was "natural" for the last 600,000 years. Remember this has to be more than any amount put into the atmosphere in the last 600,000 years of sampling because in just the last one hundred years the population has grown over 4 times and has never been much over 1 billion until the twentieth century. Every person exhales CO_2. The only conclusion you can come to is, even if all industrial greenhouse gases were eliminated, the amount of CO_2 would not be reduced below that 300ppm number. You would also have to *eliminate* five billion people. The "problem" is, if global warming is caused by people, more and more people means more and more global warming even if you stop all industrial pollution. But don't be afraid yet.

Chapter 4: The Government Solution

My dad often said "Liberals see the world the way it should be while ignoring the way it is". He also pointed out that their ideals are rarely wrong but there methods always fail. I believe he considered anyone who promotes government programs (almost always socialism by definition) a liberal. All too often we hear "the government should do something" "there ought to be a law". And when liberals and ideologues get a new law or mandate, they feel better because to them the world looks a little more like they think it should be. One example of a recent "feel good policy" is President Obama's proposal for 150 billion dollars on renewable energy research over ten years. No specifics therefore no way to gauge results! What it will do is relieve liberals of any real responsibility other than to complain. Liberals will say "at least we did something". They (especially the extreme environmentalists) will not compromise and keep trying to pretend the world can be as it should be while ignoring mankind's constant struggle with the way it is. My point to all this is, if certain groups do not see a need for a lot of compromise and thinking instead of just following an ideologue regardless of what he or she says, we cannot move forward in any direction. If you are one of these people then you deserve and can only expect more of the same.

The global doomers are typically liberal and promote government intervention into everything. They believe and want you to believe through laws, mandates and taxes any problem can be solved. (This is the same group that believes the U.S. government through "right wing conspiracies and special interests —"big money"— are the cause of the worlds' problems). I am not one who blames us or our government for the worlds' problems but I ask you to think of an issue or problem the government has solved by using laws, mandates and taxes. The best current example I can think of is Social Security. The current system can not work without extremely high taxes to our children or if a whole lot of people die young. (Social Security came to be in the 1930s when few people lived to be much past retirement age) Dying young will not be my choice and I assume not yours. So unless we get a real dialogue going and some real innovated ideas, the system is going to fail or our children will pay dearly. (Remember the train commercial where the older guy says "30 years! Why should that concern me? Social Security taxes are something that concerns me. Of course, I care about my children!) The problem is not our political system but how it is managed by our so called representatives in government. The Congress is the worst. The partisan blame game and personal attacks are a disgrace. Take ex-President Bush's recent attempt at reforming social security. Right or wrong the political attacks were vicious and he took a political hit that contributed to his decline in popularity. A great example of government failure! Reforming social security will not be addressed again until it is on the verge of total collapse. Who will risk bringing it up? Unfortunately global warming is becoming as politically dangerous.

Some proposals currently on the table are; taxing the oil companies, higher CAFE standards for cars, mandates for ethanol or other alternative fueled vehicles and lower speed limits. Wow sounds like the fabulous 1970's and the total failure of the Carter administration. (Remember President Carter campaigned on a platform that included energy independence.) Mandating fuel economy forced smaller cars to be built. Windfall taxes on oil companies caused oil shortages and higher prices. Remember gas lines and shortages. Remember the price of gas skyrocketing along with the economy tanking. The intentions might have been good but the results were and continue to be a disaster. When the economy recovered people had no stomach for government mandates that lowered their standard of living. The small cars that were produced gave the foreign manufacturers a solid and permanent share of our auto market and the loss of American union auto jobs was massive. You can argue that these jobs have returned (mostly non-union) because the foreign manufacturers assemble cars here now, but I doubt those who lost there jobs, homes and pensions would say it was a good thing. But this was done by an administration and political party that claims it is for the working man. (Sound a little like what the new administration is proposing) Only a moron can honestly believe that these people, good intentions or not, promote solutions for the working class or anyone else for that matter.

The mandates for higher fuel standards wasted years of engineering talent and vast amounts of money trying to make gasoline a good fuel for cars instead of abandoning it. The amount of pollutants coming from our tailpipes has been dramatically reduced. It all looks good, but if we had spent that time and money, through market pressures and government incentives, on alternative fuel development and distribution, we would have a different situation today. On the surface it looks like a government success story but I consider it to be the opposite.

The result was a backlash against the notion that oil supply was a major problem. Our city skies are clearer than they have been since the industrial revolution began largely because of the reduction of pollution from cars even with more and more cars on the road but our oil dependence continues to get worse. We the people felt screwed and many still do. Many think the government conspired with big oil. Many think shortages were created just to raise the price. When the price and supply stabilized, they appeared to be correct. You saw the dramatic increase in the purchase of trucks and SUVs that continue to be popular today and fuel-efficient cars were only for the poor. You also saw the decline in trust in government solutions from a people who historically have trusted their government.

The reality of the government solution was thirty years of wasted money, time and technical talent and thirty more years of dependence on foreign oil while our wealth was and is being pumped out to foreign countries with no significant long term reduction in pollutants. A total loss!

Chapter 5:   The Hypocrites

One thing you rarely see is a global doomer promoting a real and viable solution. They preach about environmental disasters and tell us how everyone else must change. They want government to legislate or mandate change. They blame big corporations and claim the big bad American corporations only look out for profits at mankind's peril. They generally are well off financially and really are not going to be effected by higher taxes or an economic slowdown. And if their stock value in the big bad corps decline they will simply sell. They won't loose their jobs when companies fail due to excessive regulations. They are lying hypocrites when they tell you they are looking out for you.

They talk about how you need to drive smaller cars or hybrids while they drive big SUVs or travel in limousines and private jets. They live in big homes and consume massive amounts of water and energy when compared to the average person.

I attended a driving safety class a few years ago. Yes it was because I had gotten a speeding ticket and picked the class instead of a conviction. The instructor, a State Trooper whose real job was accident investigation asked if anyone in the class had ever driven a nineteen fifties vintage car. I was the only one old enough to say yes. After a brief conversation about the construction of the car he pointed out the all steel dashboard and the massive weight of these cars. He then explained that in his experience of accident investigation when a big car meets a small car "the big car always wins". This has nothing to do with global warming but may give you an idea why a hypocrite may prefer a large SUV or limo. Big cars are simply safer, more comfortable and are a lot more prestigious! I think they simply don't practice what they preach and only want everyone else to "save the planet".

Be aware of these hypocrites. Do not tout them as great people because they bring attention to the problem. That is not enough! If they truly believe and are not hypocrites they must do something. When are the global doomers going to stop being hypocrites and put their money where there mouth is? I'm not talking about buying a small car, living in a small home or using public transportation, although I would no longer be able to call them hypocrites. I want to see them get their other rich liberal friends to invest all that money on real solutions. How about a start up company that spends their money on researching a viable alternative fuel! How about investing in the distribution of this fuel! How about buying Ford or GM stock, take control and do what they complain the U.S. auto industry is not doing. I for one would want them to be successful but remember if the global doomers loose their money we will be rid of them! And if they do not want to loose their money, here is another suggestion. Refuse to buy anything that uses oil, including gasoline, or any other product, including most electric power, that is made with or by burning a fossil fuel. (I'll bet you global doomers could not even buy a solar panel or a hybrid car for that matter). Come on guys and gals, do your part. Let us all say to the rich liberal global doomers, just stop complaining and do something, anything will do!

Carbon credits: I really do not know anything about carbon credits. But there can be no environmental gains. They are a non product a hypocrite can sell someone else to make money. A company or country can sell credits if they do not emit carbon to companies that do emit carbon. The hypocrites will then give these bad guys the right to exist. You will pay more for products because it will cost companies more to produce them and some hypocrite will make money! They are touted as a way to make lower emissions economically viable. True but what they mean is it is ok for them to make money by punishing those companies that can not adapt. Punishing companies that can't retool or afford to lower their emissions will only have two possible effects, loss of jobs (many of which are being outsourced to China already) and higher costs for products. Remember, carbon credits do nothing to lesson greenhouse gases and will only cost you more money to live. Reject carbon credits and the hypocrites that promote them.

Chapter 6:   The Things You Should Be Afraid Of

Everyone wants to talk about how we must stop global warming but there is no talk of reasonable plans to limit the effects of climate change. Climate change is real and constant, not new! A Fact of life! History has many examples of climate change that are indisputable. I was watching a program about the revolutionary war and George Washington crossing the ice-choked Delaware River. The crossing was immortalized in a very famous painting showing large ice flows. The Delaware River has not been ice choked in decades. A few years later, the French were starving due to a food shortage caused by you guessed it, climate change. "Let them eat cake" is, other than her demise, probably what you know most about Marie Antoinette. Yes it got cold, the growing season was shorter and there was not enough food. It will happen again. In history class, do you remember the Dark ages? Climate change! The Renaissance! Warm, fuzzy, nice times! The climate keeps changing. The science of global warming is now looking, through ice core samples of the carbon content in the atmosphere, back thousands of years. The data suggests that climate change can be correlated with the carbon content in the atmosphere and has contributed to the demise of many species. Even to some degree the dinosaurs! Guess what? Dinosaurs and even the people during the dark ages did not have cars! Getting a clue? Take note of the warm periods in history! The charts (look in Al Gores' book *An Inconvenient Truth*) show clearly that colder climate trends are much more prevalent than the warm ones. The only conclusion to come to is, our current global warming is to some degree a good thing and maybe very temporary. Scientists are also learning about how dramatic events, such as asteroid impacts, and terrestrial events like earth quakes have always changed the climate. There is no reason to believe events of this nature will not happen again. There will be many environmental disasters that will challenge our very existence. But gradual climate change is not one of them.

Be afraid of the lack of a long term environmental plan to control droughts and floods that threaten our food supply and economic viability. It's time to stop wallowing in self blame and indecisiveness and fix things!

Like I said, climate change is real and constant. More people will cause more greenhouse gases! There will be a lot more people! How many who knows! I think the United Nations estimate of population growth is very naïve. But even if it is accurate, three billion more people will cause a lot more stress on the climate and resources. Remember, three billion more people will make the total population over nine billion. That is more than nine times the total human population before the year 1800. Last century the U.S. population tripled, and it may triple again this century. That would put the U.S. population as high as one billion people in the next 100 years! We need to prepare so we can feed and cloth them.

Be afraid of government mandates on how many children to have. Do not tolerate taxing or punishing people who have more children. Can you imagine a society that allows unimpeded access to abortions (called a right to privacy) that might someday propose a limit on how many children a couple can have? Sound absurd? It's happening in China. Don't ever except a regulation or law (or even a proposal) that limits your reproductive rights! And don't you believe that fewer children will fix anything!

Be damned afraid of the notion that we the American people should pay even one dime to help stop global warming around the world. Be afraid of any treaty or agreement to reduce greenhouse gases that lower our standard of living. We don't need the Kyoto treaty. There are some people trying to convince the people of this great country that we are to blame and we should pay. I say to them, "Go to hell"! China and India are going to continue to industrialize and they are showing little concern for the environment. We can not dictate to them on how to build their economies and with their growth rate they will soon contribute more greenhouse gases and other pollutants than we ever would. Again all we can do is become energy independent with a commitment to renewable clean fuels and, most importantly, be prepared for the climate changes that are inevitable.

We have a political system that will allow fanatics to regulate our lives. We all believe in our basic form of government (although I am very discouraged about how few of us seem to understand it) and will accept stupidity, but you should be very afraid! Please think! Please learn! Please reject any notion or plan that taxes you (or even the rich) to solve a problem that is easily solved through common sense. The fanatics on both sides are going to continue to argue about whether or not global warming exists and who or what is to blame. They will do nothing or the wrong thing. That is the only thing I am afraid of! The only gloom and doom I see is not using our technology to prepare for the future. It is a bright future and like the past has many pitfalls. Ignore these pitfalls, or expect government mandates to fix them, and the future might just be gloomy.

The year 2040: Only two possibilities exist.

One we continued to fight over the existence and cause of global warming! With policies similar to the ones from the last 30 years you will see a sharp decline in the standard of living in this country. We will have taxed and regulated our industries and people to death. No jobs and no future. We will still depend on oil as our primary source of energy because we kept mandating cleaner gasoline vehicles instead of abandoning them. Our wealth will be gone because of a stupid continuation of buying foreign oil instead of pumping ours while continuing to increase government interference in our lives. With this approach our country will look more like Europe than America and we will be as insignificant as they are and socialism will have replaced the freedoms and prosperity we have had for two hundred years.

Two we started solving the real problem in 2009! Everything I will talk about in the next chapter could have been done in 1975! What I will try to demonstrate is that the technology needed has existed for 30 years or more. Or to put it bluntly very little research and "no new technology" is needed!

Chapter 7: Real Solutions And How To Get There

At this point we have three possible global warming scenarios;

One Global warming is natural and not caused by man!

The fix! Clean renewable energy, responsible environmental policies and preparation for the future through the best possible management of our resources.

Two Global warming is not natural and is caused by man!

The fix! Clean renewable energy, responsible environmental policies and preparation for the future through the best possible management of our resources.

Three Global warming does not exist!

The fix! Clean renewable energy, responsible environmental policies and preparation for the future through the best possible management of our resources.

I think you get the point. No matter what you believe, we must always look to the future and prepare for inevitable climate changes whether temporary cycles or permanent change. Waste and disregard for the planet is just stupid, immoral and suicidal.

"Put America First"! That not only sounds nationalistic, it is!

I have to believe that you cherish our way of life and particularly our standard of living. Keep in mind this all comes from our political and economic systems. Loss of wealth will mean the loss of all the things we hold dear. So how do we put America first?

I will not and I believe you will not tolerate the notion that we the American people should pay anyone to clean up pollution or reduce the world's greenhouse gas output. Not our problem! Besides if we are the biggest contributors of greenhouse gases in the atmosphere, and if we believe them to be the problem, then all we have to do is take care of our part. Make any sense to you?

What it takes to change anything is money. The place to spend that money is here. We must keep our economy strong and keep our wealth here and when it comes to energy our biggest problem is foreign oil. Drill American oil! Drill our oil if you find it in Alaska and the everglades and Senator John McCain is wrong drill it out of the Grand Canyon if it is there. The money that is put into our economy will provide jobs here, increase our tax base here and stop putting money in hands of countries that do everything they can to undermine our prosperity. With the money we stop pumping out of America, we can fund changes that make sense for America. I'm not suggesting a disregard for the environment, quite the contrary. Any environmental damage can be minimized with proper (not overbearing) laws and oversight. When it comes to off shore drilling, we are at a much higher risk of environmental damage from large super tankers spilling their cargo than we are from oil drilling platforms. Did you take note of the destroyed oil platforms during one of the recent hurricanes in the gulf? I heard nothing, because there was none, of the ecological damage due to oil spills and I believe in a few decades oil tankers, at least in America, can be a thing of the past! Would that be nice?

Don't be fooled by those that say drilling our oil will have little effect. Everyone knows we do not have enough oil to fill our energy needs but I guarantee a U.S. policy to drill our oil and move aggressively to alternatives will shake up OPEC. The cost of oil has dropped dramatically before we even scratched the ground at least in part because we, for a brief time, talked seriously about drilling here! OPEC will increase production, lower the price and make it cheaper to buy their oil than drill ours. Even at a higher cost, drill here drill now! That in fact is the pitfall! When the price of oil goes down there is no political pressure or money to drill U.S. oil. We should drill even if it raises the price of our fuel. If we do not, we will continue to be susceptible to OPEC and oil speculators and in a year or two the price of a barrel of oil will once again hurt all of us. The way to offset some of the pain of the higher cost of U.S. oil is to reduce the sales taxes at the pump to pay the additional cost of pumping our oil. T. Boone Pickens, during the time I was writing this book, started a campaign to promote American energy independence. I have not yet seen all of his ideas, but get behind him. I don't know his motives, but get behind him. I hope he is successful because he is on the right track. Drill more American oil now to keep money and jobs here. Don't mock or tear down the fundamental part of what he is saying he wants to achieve. It has to be done! I disagree with him on the use of CNG for large trucks and fleets because I think the transition to bio-diesel is easier for trucks and is more viable in the long run than CNG. I will get to that later, but notice that he has dramatically reduced his immediate plans for wind energy development. Did you ask why? The obvious answer is money! The price of oil has dropped and his projected revenues have dropped, even a billionaire has limits!

Electricity: The electricity we need is available from only a few abundant and practical American sources: Coal, nuclear, wind and solar power. Wind and solar power should be developed as fast as possible because the technology is already up to speed and they do not pollute. All we need is places to put it. But until there is a significant improvement in battery technology solar and wind can not be our primary source of electricity. Coal is another story.

Coal is a fossil fuel and puts out greenhouse gases. It is already the fuel we get most of our electricity from. Clean coal technology is being developed but you must be willing to pay higher electric bills to build the power plants. Higher electric prices will hurt the working families. But we must keep using coal, clean or not, until there is a viable American replacement. Stop using oil and gas for electricity by replacing it with coal, wind or solar as fast as possible! I am generally opposed to higher taxes but I think it is time to tax states that do not produce at least the amount of electricity they consume. It sounds harsh, but they must face the fact that if their energy is coming from far away, it costs more and wastes more. And remember much of the power grid that allows such abuse was and is subsidized by the Federal Government and belongs to us all! I could be wrong but there are trade offs. To have a high standard of living you must make some sacrifices! If you do not want polluting power plants in your state, then demand and pay for clean renewable energy with your money not mine! Nuclear power is your best bet.

Nuclear power is expensive to build but puts out no greenhouse gases. The problem, like garbage dumps, no one wants to live around one. But if you can solve the location problem and say it takes five years to build a nuclear power plant, you get twenty power companies to start building twenty plants today (use incentives not mandates) and in five years you could eliminate a lot of greenhouse gases we are concerned about and provide cheap clean electricity for years to come. No public money necessary! The power companies can and will finance a future power source such as nuclear if you get excessive regulations off of their backs. They want their business to prosper! With a growing population we will need more and more power and the only large scale *green* alternative available today is nuclear. This will take tremendous political and public support and the anti-nuke nuts are going to have to compromise. I don't hold out much hope, but it does not take thirty years to change and solve a problem, just the will to do it.

No, I did not forget hydro electric power. Hydro electric plants are green but have proven to be a poor alternative because they involve damning up rivers and streams that have serious repercussions to the ecology. I don't think we have enough rivers to provide all of our power anyway. It is ok where feasible but can not significantly help our energy problems.

The electrical grid is now a concern. It is out of date and needs extensive upgrades. I think it should be done in conjunction with new sources such as wind and solar farms, and new power plants to build a new integrated system. Let's maintain the old while designing the new!

Planes, Trains and Automobiles: A renewable transportation fuel has to be a top priority to achieve a long term solution to oil dependency. Corn-based ethanol is the one fuel that is currently being touted as a possible long term solution. But it is not being applied in a logical fashion. Right now ethanol is being added to gasoline to cut emissions. This should be stopped. The 900 million gallons (or is it billions of gallons) of ethanol mandated to be added to gasoline is being wasted and could cause a significant rise in food costs. The effect is higher gas prices with a very small reduction of oil usage and pollution. And, believe it or not, in states that mandate excessive ethanol additives to gasoline, the people pay higher gas prices then the national average! Try mandating corn based and other current bio-products be used to produce bio-diesel only. Yes bio-diesel first. Change the trucking industry, and possibly the rail and airline industries, to bio-fuel fast using all the government and industry resources necessary. The results would be lower fuel costs for shipping and transportation. This will lower the cost of gasoline because right now we use oil (mostly diesel) to transport gas. Also, if we could achieve 100 percent bio-fuels for commercial transportation, the farmers and ethanol producers (if they convert to bio-diesel) would have a permanent and stable customer base allowing that industry to profit and possibly reduce the need for some farm subsidies. Not a transfer of wealth but a transfer of priorities. You need "no new technology". The diesel engine was invented in the year 1886 to run on vegetable oil. The current fleets need little modification (unlike CNG) to run on bio-fuels so the cost would be minimal. And, because it could be a totally American source of energy, our market will control the cost and keep billions of dollars in our economy. A cheaper, clean and renewable source of fuel for transportation would help keep all industries costs down and help stabilize the economy. No fluctuations due to oil prices. And by the way, there are a number of other sources of bio-diesel besides corn and other crops. Many animal byproducts as well as some human waste products can be used to make bio-diesel. It is almost a crime now-a-days to use saturated fat (animal fat) in food or for cooking, but it makes a great fuel for your truck! I believe we can convert our diesel fleets to bio-fuels in a very short period of time. But, if we continue to use corn-based ethanol for cars, we will continue to make our problems worse while wasting an opportunity to solve a big part of the problem. Once again the government with all good intentions is prolonging the big problem when they could actually solve a big portion of the problem by stopping ethanol additive mandates and help move to bio-diesel!

Obviously, a replacement fuel for cars must be a high priority. Fuel cell cars look promising but they are very expensive and until that changes their future is limited! Electric cars are cool but have very limited use and really do not help with global warming. First their range will never be at parity with gas-powered cars. Even if you can get to a three or four hundred mile range on one charge, how do you get back? Once that is solved you still have another dilemma. You have to use the current electrical system (the Grid) to re-charge the batteries. You get a net loss in energy and the power your using still puts out greenhouse gases at least until we have green electrical sources. Electric cars will help reduce dependency on foreign oil and for that reason alone have a great place in our future. New technology, such as a standard easily swapped battery system that could be done at gas stations, could make electric cars very viable and might be the place to spend some research money. Today's electric cars are good for short trips and commuting and I hope GM is very successful with their new Volt coming out soon! I got a good look at the Volt at the Auto show in Chicago recently and was very impressed.

Imagine that! A big bad American corporation (GM) paid to develop what looks like a viable electric "green" car (The Volt) possibly at their own demise. This can't be so!

Hydrogen looks like it has tremendous potential for the future, but how long it will be before it can be produced and distributed cheaply, if ever, is anybodies guess. So we need a current technology solution if not for global warming then at least for energy independence. Hydrogen is the most abundant element in the universe and the technology for hydrogen powered cars exists. Unfortunately it is rare in its pure state and there is no current way to extract hydrogen efficiently. It is not the alternative fuel of the near future. Research is fine and important, but there are better current alternatives and tax dollars should be put towards real current solutions not dreams or speculation.

Natural gas (CNG) is relatively cheap and abundant. Yes it is a fossil fuel but happens to be the cleanest burning of fossils fuels. It also requires "no new technology" to be used as a fuel for cars. (So far this is the only disagreement I have with T. Boone Pickens) As long as the U.S. supply is abundant enough so as to not dramatically increase the cost due to an increase in demand, it could be a viable replacement for gasoline for enough time to achieve a non-fossil fuel substitute. If however, the cost to heat homes and businesses would be dramatically affected, then the working poor would be hurt. I think we have to realize the need to keep the cost of living at a minimum.

CNG does have one immediate drawback regardless of abundance and price. If you do not have gas piped into your home, you don't have a convenient place to get it. The distribution system, to my knowledge, is not there. Hopefully T. Boone Pickens has a natural gas distribution solution for cars? Yes, as he said, it may be a temporary solution, but it is American! Like I said, I disagree with him as to where to apply this temporary solution, but totally agree that it is part of a solution to oil dependency.

We would need the industry to want to invest in a distribution system to make CNG easily obtainable. The auto industry cannot be expected to build CNG-powered cars unless there is a market and that will depend on fuel being readily available. The U.S. auto makers are already in serious financial trouble. There must be at least price parity between the CNG vehicles and the gasoline vehicles (they should be cheaper because they will require fewer emission controls!). The public will not respond in a significant way unless the price is right, which is why you do not see a great demand for the current alternative fuel cars such as hybrids and flex fueled vehicles, they are just too expensive!

I don't know how difficult it would be to adapt the current gas stations to provide CNG but unless fuel is available for a fair price the idea is dead. The best way I see to do it would be to form a partnership between the CNG suppliers and the auto dealers. If the auto dealers at least temporarily carried CNG fuel, the consumers would have a place to get it until the distribution system is up to the level of the gasoline distribution system. Auto dealers generally have room for temporary filling equipment. There are safety considerations but we already have propane filling tanks at hardware stores, home centers, convenience stores and gas stations with few incidents. It can be done!

The oil and natural gas industry has a large stake in it and the money to fund a distribution system. With some government incentive and possibly some slack in regulation, all they need is a little time and a consumer commitment. Those of us with natural gas in our homes could get the gas through a home installed pumping system. This equipment already exists. Again, "no new technology" needed! If as abundant as they say, CNG could replace 90 percent of gasoline for "new" cars in less than five years with just a small effort. In ten years, because the older gasoline powered cars would become warn out and obsolete, you could see gasoline almost as hard to find as ethanol is today because more new and used CNG powered vehicles (as well as other alternative fuels) are available. And we would not care! I think ethanol/electric hybrids, CNG/ethanol flex fueled and all electric cars are going to be the choices of the future. To do it will take government incentives for the consumers as well as the auto and gas industries. The price of oil would come down the moment you started an aggressive substitute plan because oil speculators would stop bidding the future price of oil higher and higher because the financial risk would be too great. Lower fuel costs, less dependence on foreign oil, lower emissions and more money in the U.S. economy would be an immediate result. All good!

Home Heating Oil: Home heating oil is a source of concern for millions of American families. I will never heat with oil again. I hope the bio-fuel industry can produce enough bio-diesel to help with this problem. One way it can help, even if there is not enough bio-fuel for home heating, is the cost of heating oil (which is diesel fuel) would be dramatically reduced if we moved to bio-diesel for the trucking industry, because of a tremendous increase in supply and a decrease in the transportation costs. This would be an immediate result of bio-fueled diesel fleets! And, as soon as there is a substitute fuel for cars, the supply would be even greater and therefore I think oil would be even more affordable! The cost has to be brought down because it will be decades before home heating oil can be phased out. The working class simply can not afford new heating systems. I think there will be a need to stockpile home heating oil, similar to our strategic oil reserves, to maintain the supply and therefore the price. In a few decades U.S. oil and gasoline might even be an export at a fair price that is not influenced by OPEC rather than an import and sold to friendly countries that have no access to bio-fuels.

A Lay Persons Note about Crude Oil: The way I understand crude oil is that when it is refined there a four or five stages of refinement that produce different products. First, is the heavy tars used for asphalt and similar products, then there are a couple of other stages that are used for materials for plastics and heavy lubricants and the higher stages are the diesel fuels and gasoline. (I don't know the exact details but, oil is always going to be needed for the production of many products. That is another reason to stop wasting it in our gas tanks.) I believe that the fuels are always going to be made because they are a result of the refining process, unless they can be reprocessed for other products. You and I do not need to research this – the refining industry already knows! And the American oil companies are in no danger of becoming obsolete or un-profitable. They will adapt and be fine! It is time to stop demonizing the oil industry. It has helped create the high standard of living we have and we need them in good health to help achieve our alternative energy goals. The U.S. oil reserves can be a source of domestic prosperity for centuries to come simply by using it only where needed.

Water is it! No, I'm not suggesting we can use water to fuel our cars. Water is mostly hydrogen and maybe the main source of hydrogen in the future, I don't know. But it is not a fuel source per say. Water however is the most important resource we have. I'll get to the point. The general consensus is that climate change is happening. Global doomers say it is our fault and the result will be droughts and floods and possibly the end of mankind. Droughts are not enough water! Floods are too much water! I know that did not need an explanation, but I could not resist.

My wife constantly points out (complains if you will) that she has to flush the toilets multiple times. I try to explain that the new water saving systems don't produce the amount of flow that the old toilets did. She doesn't need to know how a toilet works but because she is an intelligent person, she recognizes a system that does not work properly. I explained that these new toilets are mandated by the building codes to preserve water. Her comment, "well it doesn't work". Apparently my wife is not the only one aware of it. A friend of mine has figured out that you can modify them and they will flush better. Of course, all he is doing is increasing the amount of water used. A well known conservative commentator mentioned liberal government bureaucrats had made toilets very inefficient and now need multiple flushes. It is sad that toilets have risen to such a high level in our social concerns but it points out that stupid government mandates always fail to achieve their goals and always cost us money. Conserving water is certainly a good thing but flushing multiple times wastes a lot more water than the old toilets did and common sense approaches by government are rare.

Water is the resource we should be concerned with most and although the amount of water on the planet is relatively constant, the distribution of it is constantly "naturally" changing due to (you guessed it) climate change.

I propose a continental water control and supply system. Yes there are a whole lot of things to be considered that will involve all the political, social and even engineering communities to work together, but before the fighting starts, consider the following ideas.

The oceans are rising! Build desalination plants (preferably nuclear powered built in concert with the power plants and an improved grid) along the Pacific, Atlantic and Gulf coasts so there never has to be any area concerned about not enough water. "No new technology" needed!

The USS Ronald Reagan (I believe our newest nuclear powered aircraft carrier) produces all the fresh water it needs to stay at sea indefinitely and will only need re-fueling once in its life! It does not emit greenhouse gases and has a crew of over 5,000 people. They do have to have a source of food. What does that tell you? One, is we can get all the water we need! Two, even the USS Ronald Reagan like our vast grain belts are not big enough to provide food for its crew, or in our case, future generations! (Another, unrelated note, the U.S. Navy has used nuclear power safely for 60 years. Tell that to your liberal, anti-nuke friends.)

Build reservoirs some lakes, some underground tanks, not big like Lake Meade, but small plentiful ones. Build pumping stations and systems of pipes similar to the Alaskan oil pipe line. Use the Interstate and other highway right of ways, where possible, for pipe lines to avoid disrupting existing communities with new easement problems. Be able to move water from areas that have too much, for example the Mississippi in the spring, to storage tanks, reservoirs or even out to sea when necessary. Have large quantities locally available to irrigate the vast farms that provide our food. I am not talking about damning up rivers to build large reservoirs. I'm talking about a lot of small reservoirs that are part of an interconnecting continental system to help nature maintain lakes, rivers and streams, as well as provide water where and when it is needed. That is what I think the systems capability should be. We should start to prepare for the inevitable climate change the global doomers say we caused.

What are the goals? One goal should be to provide water to maintain our current farm system. Without plentiful inexpensive food there is no future. How stupid would we be if we got to a point where we are energy independent and squandered our food independence do to poor planning and no insight into the future. Water is fundamental and climate change such as droughts have happened before and will happen again. It will not even take a cataclysmic drought to show us how vulnerable our food supply is. The Northeast just got through a very unusual drought a few years ago. The Southeast is just now at what we hope is the end of a great drought. The Texas panhandle is in the mist of drought and central California's vast fruit and vegetable farms (not to mention their cities) almost totally dependant on irrigation may not have enough water this year.

Another goal should be to provide water for our communities. The result would be to minimize the effect of regional droughts and flooding. We should not accept the fact that even in the Great Lakes area a growing population will eventually tax that great resource. And certainly, clean abundant drinking water must always be a high priority. In the most prosperous nation on the planet, how can we continue to ignore such a fundamental issue?

What should the main goal be? Irrigate the vast southwestern U.S. to build the energy capital of the world. If we turn the deserts into vast farms guess what we should grow. Sugar cane and other energy rich crops! We make ethanol from sugar, which by the way is a better and cheaper source of ethanol than corn and we don't risk our food supply. Much of the land is already federally owned. The populations are sparse. These vast sugar farms are also the perfect place for huge wind and solar farms. In fact, they could be totally integrated and be an engineering marvel. You don't have to cut down trees that help clean our air, like is happening to the Amazon. You don't have to risk large bird or animal populations, one of the issues surrounding some areas. And of course we must preserve some desert areas.

The infrastructure that is needed to build and maintain large solar and wind farms can be viable if there are a variety of assets to build strong permanent communities. If climate change causes a cooling of the northern hemisphere and therefore shorter growing seasons, you will have an alternate place to grow food. You might even find the start of a solution to the immigration issue. You certainly will need a large labor force to achieve these goals! The system can even include northern Mexico (if they want) and provide economic viability for Mexicans. If you are successful with that, I don't know where we will get a large labor force for our economy. I hope that is the worst of our problems!

It would give us the ability to phase out fossil fuels (such as gasoline and maybe even CNG) for our cars and who knows maybe enough surplus to provide clean renewable fuels for parts of the world that are being strangled by OPEC or have no bio-resources. Big irrigated farms and surplus water for re-forestation could also help with the reduction of CO_2 (maybe even help get us below that magic 300ppm) because all plants absorb CO_2. To put it another way, the population of the planet includes plants and animals. They provide a "natural" CO_2 offsetting system. As the people/animal population increases we should increase the plant population. Water, whether from rain or irrigation, has to be available to do this!

Another thing that has potential is a piped water system would use power to pump the water to higher elevations. Once the water is flowing down, put small hydro-generators on the system to help offset the power used to pump the water up. Yes there is a net energy loss but with a combination of gravity fed (man-made, green and no ecological disruptions) hydro generators and solar panels you might get a net gain! Every little bit will help! The engineering skills and most of the geological information already exist, we don't have to wait and "no new technology" is needed!

It would be a great achievement. It would create permanent construction and civil works jobs. It would have to be funded on a long term basis similar to the Interstate Highway System and take years to build but it also would create a long term vision for America's future, something we desperately need. Like the Interstate Highway System it would create new areas of prosperity for future generations, such as resort communities, more viable farm communities and who knows, maybe a whole new manufacturing economy based on the needs of the local population and the infrastructure to support it. Remember many opponents of the Interstate Highway System believed it was an un-necessary expense and could not have imagined the economic and social gains that have proved to be true. The Native American reservations of the southwest could see economic possibilities (beyond casino's) that would create wealth and a future for their people, building viable communities that bring their standard of living to parity with the rest of the nation without sacrificing their culture. All things considered however, a permanent viable economy would be the right thing for everyone. New, growing communities can attract workers from economically depressed areas and maybe help slowdown urban sprawl and attract some of the working poor away from urban areas to places that have a potential for "the American Dream".

The political problems are minimal, except for the extreme environmentalists. But even they can not deny that our society can not continue without trade offs. Responsible development and preservation of the resources we have is the only way to a secure future for us and all of the other inhabitants of this planet. Achieving local renewable recourses that do not damage the environment is their goal and that, is the one thing, without a doubt, they are right about. With proper planning and engineering of an integrated water and energy system, you could see a surplus of electrical and bio-energy that is American and clean with almost no environmental hazards. Who knows, maybe in thirty years or so we would be the source of much of the worlds' clean, renewable energy. Maybe the Saudis' will buy it?

I also see a day, once we get over being a throw away society and truly get committed to recycling and renewable resources, that we in America dig up our vast strategic material reserves. I'm trying to be funny! What I am talking about is, with a greater demand for materials and an advance in technology, it may be economical to dig up all the landfills to regain the waste of the past century. Wouldn't that be something?

A quote from the Intergovernmental Panel on Climate change, which I found in a text book called "Earth Science" ninth edition by Edward J. Tarbuck and Frederick K. Lutgens Copyright 2000,1997 Prentice Hall;

"Uncertainty does not mean that a nation or the world community cannot position itself better to cope with the broad range of possible climate changes or protect against potentially costly future outcomes. Delaying such measures may leave a nation or the world poorly prepared to deal with adverse changes and may increase the possibility of irreversible or very costly consequences. Options for adapting to change or mitigating change that can be justified for other reasons today (e.g., abatement of air and water pollution) and make society more flexible or resilient to anticipated adverse effects of climate change appear particularly desirable."

I believe securing our food supply and creating long term infrastructure jobs can be "justified" today. I also think climate change is inevitable therefore being prepared for it is "desirable".

Conclusion

One thing for certain is we have to stop being sheep and demand a real commonsense approach to our future that includes all Americans. Remember, when you mandate and tax anything, you raise the cost of living and screw the working class. Put a stop to the continuing story of whose fault everything is so we have someone to blame and an excuse for failure. Only "We the People" can do it!

Is putting America first wrong or crazy? I expect every country to put their people first. That is the primary responsibility of government. Don't you agree? If we use 30 percent of the energy and produce 30 percent of the greenhouse gases then we only have to take care of ourselves and we solve 30 percent of the global warming problem. We as Americans can not dictate to the rest of the world on what, how or whether to solve their energy and resource problems but we can show them the way. We do not need and should not join the Kyoto treaty. We should stay totally independent and set a course to responsible stewardship of our country. All we have to do is produce clean renewable American energy and properly manage our resources. If we are successful, we will be strong and prosperous to help, as we always do, those that need and want help!

You may think and I am sure this could be the most outlandish idea in recent times. Imagine some idiot proposing pumping water into a desert and trying to grow food and energy crops. (I must remind you that a large portion of California's fruit and vegetable farms rely totally, and successfully I might add, on irrigation!) I guarantee I'm mentally off balance at least a little! But think back. Imagine if you or I suggested in the late 1940's we could cure polio! Imagine if you or I had proposed an interstate highway system in the 1950's. Imagine what they would call you if you proposed putting a man on the moon in 1960. The names you or I would have been called don't have to be imagined they are already on the tip of your tongue. In our current political climate, I don't know where we are going to find great leaders like we have had in the past. Remember the powers that be and the extremists will reject and criticize any new thinking with the hope of killing it because un-solved problems, real or perceived, are the source of most of their political and social power. We don't have to let this situation continue.

Your part has to be a rejection of anyone too committed to blaming, with no intention of helping. Compromise will be necessary! Working with the people you disagree with most will be the challenge. We the people should reject anyone who will not contribute in good faith! Then, the year 2040 will be a time when we look forward to new challenges and not be talking (like we are now) about why something was not done 30 years ago!

If you read this book the first miracle happened. It got published! If I helped anyone see a brighter future, as I do, then a second miracle happened. If the thoughts and opinions I expressed are totally wrong, but it gives better ideas to the more intelligent, I've done my part. But most important, I hope you all start to demand a bright future through better cooperation and use of our resources so the future of this great country can be better than ever imagined. Then there is nothing to be afraid of! Thank you!

REFERENCES

All statistics and references are based on my cumulative memory and have not come from any specific source unless otherwise stated. Therefore they are to make a point and not quote or support any scientific data and are not directly related to the conclusions or opinions of this book. Just a way to show anyone including you can come to a conclusion with a little thought.

EARTH SCIENCE – Ninth Edition By Tarbuck – Lutgens

ABOUT THE AUTHOR

I have always been a working man. I am currently a partner in a small manufacturing company and although I have many managerial responsibilities, my primary function is technical design and support. Since I have been in the industry I have always had the responsibility to correct, update and write technical manuals with only the knowledge of the product to work with. So I am not new to writing and formatting text but instruction manuals are quite different than trying to express thoughts and points of view. When I was young and trying to set a course for my future it was not even considered, but now 1 believe a writing career might have suited me. Although a challenge, it may turn out to be one of the most enjoyable experiences of my life. I hope you have enjoyed or at least have gotten something of value from reading this book. Again, I thank you!